Kids Easy Readers
921 ALMEIDA
Slade, Suzanne
June Almeida
33410017252521   09-23-2022

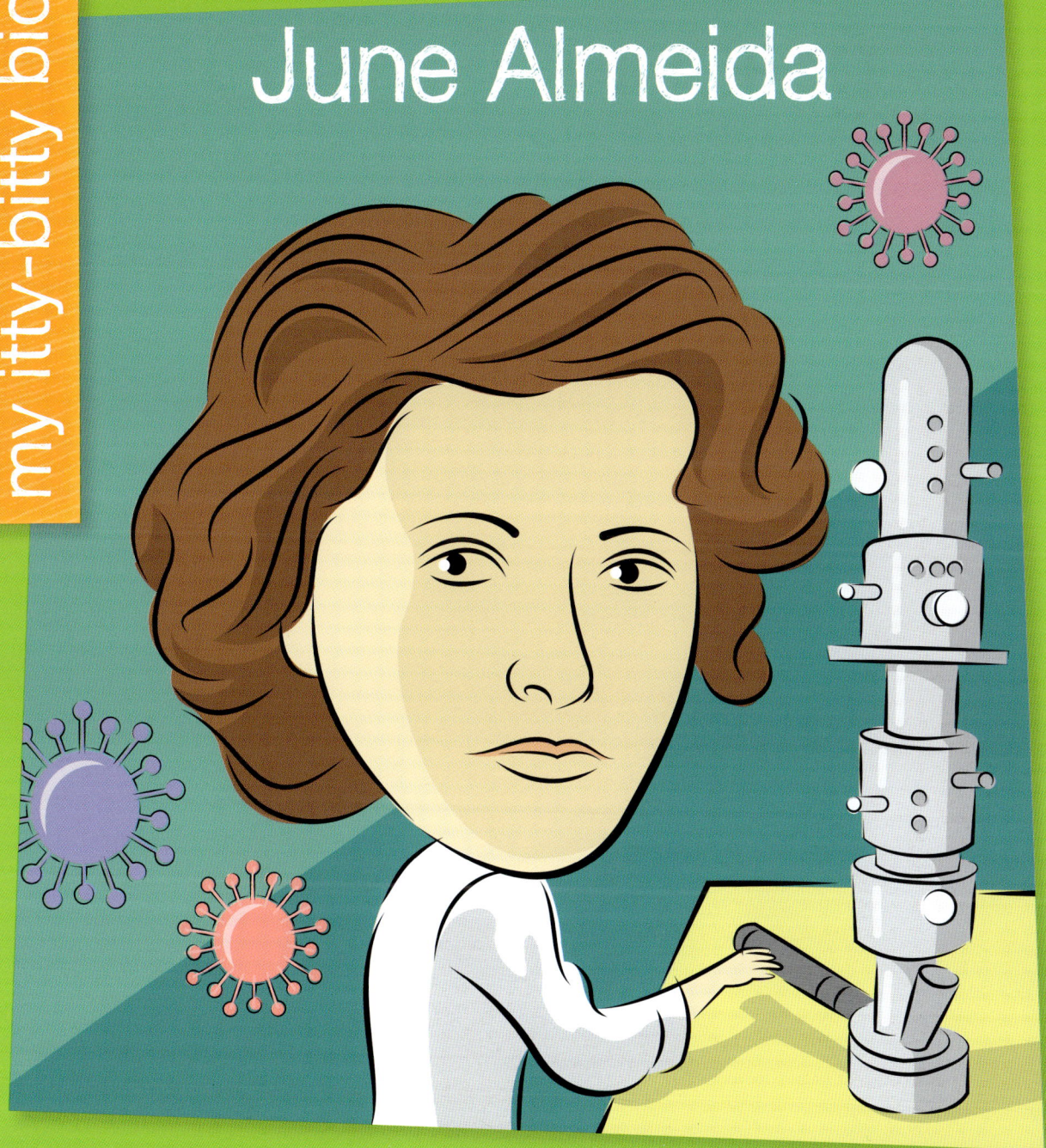

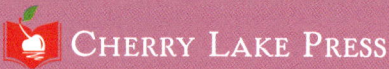

Published in the United States of America by Cherry Lake Publishing Group
Ann Arbor, Michigan
www.cherrylakepublishing.com

Reading Adviser: Beth Walker Gambro, MS, Ed., Reading Consultant, Yorkville, IL
Book Designer: Jennifer Wahi
Illustrator: Jeff Bane

Photo Credits: ©Anete Kaleja/Shutterstock, 5; ©Joyce Almeida, 7, 9, 13, 15, 22; ©WDH59510 at English Wikipedia/Wikimedia, 11; ©Photo by Paolo Monti/Wikimedia, 17; ©NIH Image Gallery/National Institute of Allergy and Infectious Diseases/flickr, 19, 23; ©Everett Collection/Shutterstock, 21; Cover, 1, 10, 12, 18; Various frames throughout, Shutterstock

Copyright ©2022 by Cherry Lake Publishing Group
All rights reserved. No part of this book may be reproduced or utilized in any form or by any means without written permission from the publisher.

Cherry Lake Press is an imprint of Cherry Lake Publishing Group.

Library of Congress Cataloging-in-Publication Data

Names: Slade, Suzanne, author. | Bane, Jeff, 1957- illustrator.
Title: June Almeida / by Suzanne Slade ; illustrated by Jeff Bane.
Description: Ann Arbor, Michigan : Cherry Lake Publishing, [2022] | Series: My itty-bitty bio | Includes index.
Identifiers: LCCN 2021007984 (print) | LCCN 2021007985 (ebook) | ISBN 9781534186842 (hardcover) | ISBN 9781534188242 (paperback) | ISBN 9781534189645 (pdf) | ISBN 9781534191044 (ebook)
Subjects: LCSH: Almeida, June D.--Juvenile literature. | Scientists--Juvenile literature. | Scientists--England--London--Biography--Juvenile literature. | Coronaviruses--Juvenile literature.
Classification: LCC Q143.A46 S526 2021 (print) | LCC Q143.A46 (ebook) | DDC 579.2092 [B]--dc23
LC record available at https://lccn.loc.gov/2021007984
LC ebook record available at https://lccn.loc.gov/2021007985

Printed in the United States of America
Corporate Graphics

table of contents

My Story 4

Timeline 22

Glossary 24

Index 24

About the author: Suzanne Slade loves science. She studied engineering in college and worked on cars and rockets. Now she's an author and has written more than 100 children's books. Many are about science. She also wrote a book called *June Almeida, Virus Detective! The Woman Who Discovered the First Human Coronavirus*.

About the illustrator: Jeff Bane and his two business partners own a studio along the American River in Folsom, California, home of the 1849 Gold Rush. When Jeff's not sketching or illustrating for clients, he's either swimming or kayaking in the river to relax.

my story

I was born in Scotland in 1930.

I loved my family. We lived in a **flat**.

I liked to learn. Science was my favorite subject. I was smart.

I won a science award at school.

What subjects do you like?

I had a camera.

I learned to take beautiful pictures.

I wanted to go to college. But my family could not pay for it.

I left school at 16.

I got a job in a hospital lab.

I used a **microscope**. It made images of **cells** large.

I took a new job. The lab had a bigger microscope.

I did not know how it worked. But I learned.

What have you learned?

This microscope could take pictures. I took pictures of **viruses**.

Doctors used the pictures to learn about **diseases**.

My work helped sick people.

I discovered a new type of virus in 1966.

It is called "**corona**" because of its crown shape.

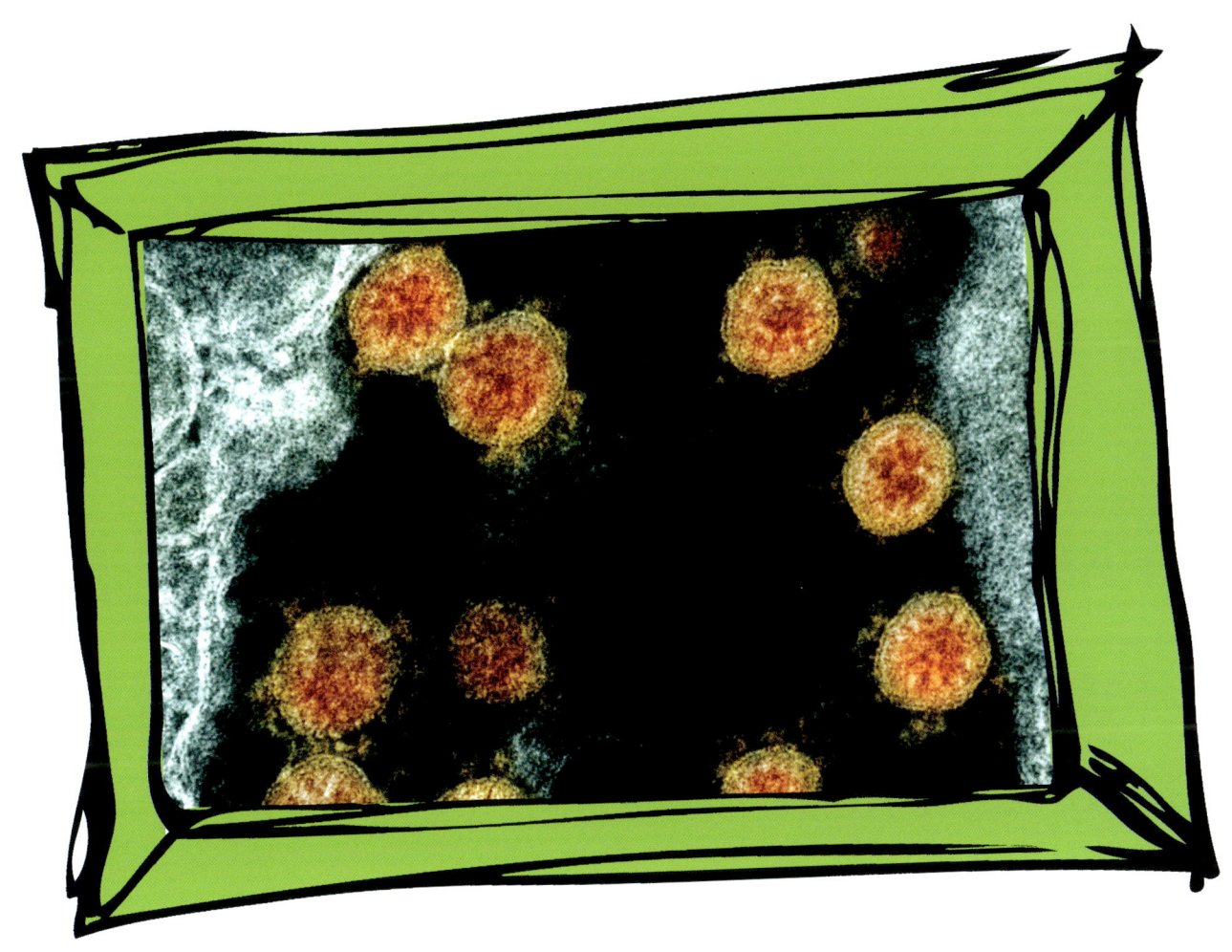

I kept learning all my life.

I taught myself to play the flute.
I liked **yoga**.

I died in 2007. My work and pictures still help people today.

What would you like to ask me?

timeline

1947

1930

● Born 1930

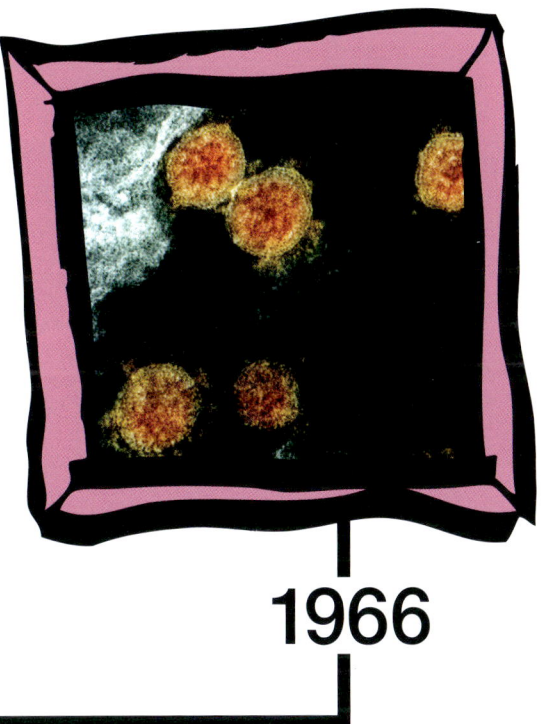

1966

2030

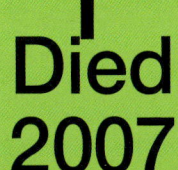
Died
2007

glossary

cells (SELS) the smallest structural units of an organism

corona (kuh-ROH-nuh) a shape similar to a crown; a severe, contagious virus is called the coronavirus

diseases (duh-ZEEZ-ez) harmful illnesses

flat (FLAT) an apartment in a big building

microscope (MYE-kruh-skohp) an instrument for making smaller objects look larger

viruses (vye-RUH-sez) small particles that can make people sick

yoga (YOH-guh) breathing exercises and body movements that help you feel better

index

cells, 12
corona, 18

family, 4, 10
flat, 4

microscope, 12, 14, 16

pictures, 8, 16, 20

school, 6, 10
science, 6
Scotland, 4

virus, 16, 18